John Reinhold Forster

A Catalogue of the Animals of North America

Elibron Classics
www.elibron.com

Elibron Classics series.

ISBN 0-543-70093-3 (paperback)
ISBN 0-543-70092-5 (hardcover)

This Elibron Classics Replica Edition is an unabridged facsimile of the edition published in 1771 by B. White, London.

A

CATALOGUE

OF THE

ANIMALS

OF

NORTH AMERICA.

CONTAINING,

An Enumeration of the known QUADRUPEDS, BIRDS, REPTILES, FISH, INSECTS, Crustaceous and Testaceous ANIMALS; many of which are New, and never described before.

To which are added,

SHORT DIRECTIONS

FOR

COLLECTING, PRESERVING, and TRANSPORTING, all Kinds of

NATURAL HISTORY CURIOSITIES.

By JOHN REINHOLD FORSTER, F. A. S.

Non ad unam Natura formam opus suum præstat; sed in ipsa varietate se jactat.

SENECA, Quæst. Nat. VII. 27.

LONDON:

Sold by B. WHITE, at Horace's Head, in Fleet-Street.

M.DCC.LXXI.

TO

MARMADUKE TUNSTALL, Esq.

Dear SIR,

THE repeated favours *You* were so kind to bestow upon me, in the compass of a short acquaintance, and the zeal with which *You* promote the great cause of Natural History; encouraged me to prefix *Your* Name to this Publication, and give me an opportunity, thus publicly, to acknowledge the assistance *Your* benevolent and generous friendship favoured me with.

Nothing is left to me, but the simple mention of thanks; and my sincere wishes for *Your* health, happiness, and the enjoyment of every intellectual and moral blessing. I am, with the truest regard,

Dear SIR,

your most obliged,

obedient,

humble Servant,

London, April 24, 1771.

JOHN REINHOLD FORSTER

To the READER.

I Had hinted in the Preface to the third volume of *Kalm's* Travels, that I could publiſh but an imperfect and ſmall catalogue of *North American* animals; and therefore declined to give it. Since that time, I have been preſſed by ſome worthy friends to publiſh that catalogue, ſuch as it is; and what is ſtill more, I have been favoured with ample materials by a Gentleman, who is forming a collection for a Natural Hiſtory of *North America*, and hopes by this to incite the inquiſitive and learned reſident in that country, to ſearch, and to tranſmit to their friends in *England*, the productions of their ſeveral provinces. The Zoology of the firſt four claſſes of animals in *Great Britain*, has been very accurately and compleatly publiſhed; that of the country of the deſcendants of *Great-Britain* ought with moſt propriety to follow, as it intereſts the Mother Country moſt. Theſe reaſons had a great weight with me; and I offer this ſmall catalogue merely as an eſſay towards forming a more compleat Natural Hiſtory of that extenſive continent. To inſtruct the Collectors, I have added to this liſt ſome ſhort directions for the beſt method of preſerving and tranſporting the various ſubjects of Natural Hiſtory.

The Quadrupeds of this liſt are referred to the Synopſis of Quadrupeds of Mr. *Pennant*; the Birds, Fiſh, Reptiles, Inſects, and Cruſtacea, to *Linnæus*, *Briſſon*, Mr. *Edwards*, and

and *Catesby*; some few excepted, which are referred to the *Planches enluminées*, published at *Paris*, and marked here *Pl. enl.* so as *Catesby* with a single *C.* and those which are common to *Great Britain* and *America*, to the *British* Zoology.

The Animals which have recently been discovered in *North America*, or overlooked by Mr. *Catesby*, are distinguished by *N. S.* marking a *New Species*; and by *B.* and *Mus. Bl.* The first authority is from a Collection formed by a Gentleman in his voyage to *Newfoundland*; the second, from a most select and numerous Collection of *American* Animals, belonging to a Lady in *Lancashire*.

The New Species, in the Catalogue of Birds, I had leave to insert here from Mr. *Pennant*'s manuscript; and those of Insects are taken from my own manuscript descriptions of such Insects as were not described by Dr. *Linnæus*.

The print prefixed here represents an elegant little Falcon, drawn and engraved from a fine specimen lately brought over from *North America*.

N. B. *The Animals marked* E *are of* European *Origin*; *having been introduced there by the* Spaniards *or* English, *after the discovery of the New World: and those marked* Eur. *are originally natives of both Continents.*

CLASS I. QUADRUPEDS.

DIV. I. Hoofed.

SECT. I. Whole Hoofed.

Genus.			ſyn. quad.
I. Horse	Generous	*E.*	N° 1
	Aſs	*E.*	3

SECT. II. Cloven Hoofed.

II. Sheep	Common	*E.*	8
III. Goat	Domeſtic	*E.*	p. 14
IV. Deer	Elk, or Mooſe,	*Eur.*	N° 35
	Rein	*Eur.*	36
	Stag	*Eur.*	38
	Virginian	N. S.	39
	Roe?		43
V. Hog	Common	*E.*	54

DIV. II. Digitated Quadrupeds.

VI. Dog	Faithful	*E.*	N° 110
	Wolf	*Eur.*	111
	Fox	*Eur.*	112
	Arctic	*Eur.*	113
	Grey		114
	Silvery		115
VII. Cat	Brown		129
	Common	*E.*	133
	Lynx	*Eur.*	135
	Bay Lynx	N. S.	136

VIII. BEAR	Black	*Eur.*	138
	Polar		139
	Wolverene	*Eur.*	140
	Raccoon		141
IX. BADGER	American	N. S?	143
X. OPOSSUM	Virginian		144
XI. WEESEL	Stoat	*Eur.*	N° 151
	Pine Martin	*Eur.*	155
	Fisher	N. S.	157
	Pekan and Vison		
	Striated		166
	Skunk		167
XII. OTTER	Greater	*Eur.*	173
	Lesser	*Eur.*	174
	Sea		175
XIII. HARE	Common	*Eur.*	183
	Alpine	*Eur.*	181
	Rabbet	*E.*	185
XIV. BEAVER	Castor	*Eur.*	190
	Musk		191
XV. PORCUPINE	Canada		196
XVI. MARMOT	Maryland		198
	Quebec	N. S.	199
XVII. SQUIRREL	Common	*Eur.*	206
	α Hudson's Bay		
	Grey		209
	Black		210
	β Cat		
	Ground		216
	Flying	*Eur.*	221
	β hooded		
XVIII. RAT	Black	*E.*	226
	Brown ?		227
	Water	*Eur.*	228
	Mouse	*E.*	229

	Field	*Eur.*	230
	Short-tailed	*Eur.*	233
XIX. Mole	β. yellow		p. 312
	Radiated		Nº 243
	Long-tailed	N. S.	244
	Brown		245
XX. Walrus	Arctic		Nº 263
XXI. Seal	Common	*Eur.*	265
	Great	*Eur.*	266
	Rough	N. S.	267
	Hooded	N. S.	268
	Harp	N. S.	269
	Little	N. S.	270
	Ursine		271
	Leonine		272
XXII. Manati	Manati		273
XXIII. Bat	New York	N. S.	283
	Long haired	N. S.	p. 367

CLASS II. BIRDS.

DIV. I. Land Birds.

SECT. I. Rapacious.

I. Vulture.	Carrion. V. aura. 122. C. I. 6.	
II. Falcon	Bald Eagle	F. leucocephalus. C. I. 1.
	Sea	F. ossifragus. in exhibitions
	Ringtail	F. Fulvus. *Edw.* I.
	Black-bellied	N. S.
	White	*Du Pratz*, II. 75.
	Osprey	F. Haliaëtus. C. I. 2.

Gentil-

	Gentil Falcon	F. gentilis *Mr. B*
	Goſhawk	F. palumbarius. *Lawſon*
	Aſhcoloured	*Edw.* 53.
	Sacre	*Mr. B. de Buffon*
	Winter	N. S. *Muſ. Bl.*
	Duſky	*Edw.* 3. 4.
	Rough-footed	N. S. *Mr. B.*
	Chocolate	N. S. *do.*
	Marſh	*Edw.* 291.
	Buzzard	F. Buteo. *Mr. B.*
	White-rump'd	F. Hudſonius. *Edw.* 107.
	Fork-tail	F. furcatus. C. I. 4.
	Pigeon	F. columbarius. C. I. 3.
	Little	F. ſparverius. C. I. 5.
	*	
III. Owl	Great	Strix bubo. *Edw.* 60
	Short-eared	*Br. Zool. Muſ. Bl.*
	Red	St. Aſio C. I. 7.
	Mottled	N. S. *Muſ. Bl.*
	* *	
	Snowy	St. nyctea. *Edw.* 61
	Barred	N. S. *Muſ. Bl.*
	Canada	*Br.* I. 518.
	Brown	*Br. Zool. Mr. B.*
	White	*ib.*
	Hawk	*Edw.* 62.
	Little.	*Muſ. Bl. Br. Zool.*
IV. Shrike	Cinereous.	Lanius excubitor. C. *app.*
	Creſted.	L. Canadenſis
	Louiſiane.	*Pl. enl.* 397.

SECT. II. Pies.

V. Parrot	Caroline.	Pſittacus Carolinenſis. C. I. 11.
	Illinois.	Pſ. pertinax. *Edw.* 234. *Br.*
VI. Crow	Raven.	*Br. Zool. Mr. B.*
	Carrion.	*B. Zool. Muſ. Bl.*
	Magpie	*Br. Zool. Edw.* p. 126.
	Cinereous	C. Canadenſis
	Blue.	C. criſtatus. C. I. 15.

VII. ORIOLE	Redwing	Oriolus Phœniceus.	C. I. 13
	Baltimore	O. Baltimorus	48
	Spurious	O. Spurius	49
	Brown-headed	N. S. *Mr Kuckahn*	
VIII. GRACULE	Shining	Gracula quiſcula	C. I. 12
IX. CUCKOO	Caroline	Cuculus Americanus.	C. I. 9.
X. WOODPECKER	White-billed.	Picus principalis	C. I. 16
	Buff-cheeked	P. pileatus	C. I. 17
	Gold-wing	P auratus	18
	Scarlet	P. erythrocephalus	20
	Red-bellied	P. Carolinus	19
	Spotted	*Br. Zool.* I. 178. *Muſ. Bl.*	
	Hairy	P. villoſus	19
	Yellow-bellied	P. varius	21
	Little	P. pubeſcens	*ib.*
	Black	*Albin*	
	Three-toed	P. tridactylus. *Edw.*	
XI. NUTHATCH.	Greater	*Br. Zool.*	22
	Leſſer		*ib.*
	Canada	*Br.* III. 593	
XII. KINGFISHER	Great	Alcedo alcyon	69
	Louiſiane	*Pratz.* II. 82	
XIII. CREEPER	Bahama	Certhia flaveola	59
	Common	*Br. Zool. Muſ. Bl.*	
XIV. HUMMER	Red-throated	Trochilus colubris	C. I. 65

SECT III. GALLINACEOUS.

XV. TURKY	Wild	Meleagris Gallopavo.	
	†		
XVI. GROUS	Ruffed	Tetrao umbellus	*Edw.* 248
	Pinnated	T. cupido	C. III. 1.
	Long-tailed	T. Phaſianellus	*Edw.* 117
	Spotted	T. Canadenſis	*Edw.* 71, III.
	White	T. Lagopus	*Edw.* 72
	† †		
	Partridge	T. Virginianus	C. III. 12
	Quail ?	T. Mexicanus	

SECT.

SECT. IV. Columbine.

XVII. Pigeon	Paſſenger	Columba migratoria	C. I.	23
	Caroline	C. Carolinenſis		24
	White-crowned	C. leucocephala		25
	Canada	C. canadenſis		
	Ground	C. Paſſerina		24

SECT. V. Passerine.

XVIII. Thrush	Mimic	Turdus polyglottos		27
	Ferruginous	T. rufus		28
	Fieldfare	T. migratorius		29
	Red-leg'd	T. plumbeus		30
	Black	*Br.* II.		245
	Little	*Edw.* 296	C. I.	31
	Golden-crowned	Motacilla auro-capilla	*Edw.*	252
XIX. Stare	Creſcent	Alauda magna	C. I.	33
	New York	N. S. *Muſ. Bl.*		
XX. Chatterer	Prib.	Ampelis garrulus. C. I. 46. *Edw.*		
XXI. Grosbeak.	Croſsbill	*Br. Zool.*	*Edw.*	303
	Pine	*Br. Zool.* IV.	*Edw.*	123, 124
	Creſted	Loxia cardinalis	C. I.	38
	Mexican	L. Mexicana	*Muſ. Bl.*	
	Roſeate	L. Ludoviciana		
	Spotted	N. S. *Muſ. Bl.*		
	Miſſiſipi			
	Dominican	L. Dominicana		
	Canada	L. Canadenſis		
	Black	L. nigra	C. I.	68
	Blue	L. cærulea		39
	Violet	L. violacea		40
	Fantail	*Pl. enl.*		380
	Grey	*Pl. enl.*		393
	Red-headed	N. S. *Mr. Kuckahn*		
XXII. Bunting	Snow	Emberiza nivalis	*Edw.*	126
	Winter	E. hyemalis	C. I.	36
	Rice	E. oryzivora C. I. 14.	*Edw.*	291

	Painted	E. Ciris C. 44 *Edw.* 130. 273	
	Louifiane	E. Ludoviciana.	
	Blue ?	*Br.* III.	298
XXIII. Tanagre	Blue	Tanagra cyanea	C. I. 45
	Green	Motacilla guira *Edw* 351. *Muf Bl.*	
	Bifhop	T. Epifcopus. Du Pratz. II.	94
	Red	T. rubra	C. I. 56
XXIV. Finch	Towhe	Fringilla Erythrophthalma	34
	Golden	Fr. triftis	43
	Orange	Fr. zena	42
	Mountain	Fr. montana	*Edw.* 269
	Little	C. I. 35	
	Cow-pen	34	
	Bahama	Fr. bicolor 37	
	Caroline	*Pl. en.* 181. 223 ?	
	White-throated	*Edw.* 304	
	Ferruginous	*Edw.* 354. *Br.* III. 296	
	Crimfon head	N. S. *Muf. Bl.*	
	Fafciated	N. S. *Do.*	
	Greater red-poll.	*Br. Zool.* *Muf. Bl.*	
XXV. Fly-catcher	Tyrant	Lanius tyrannus	C. I. 55
	Fork tail	Mufcicapa tyrannus	
	Chattering	C. I. 50	
	Crefted	Mufc. crinita C. I. 52	
	Black-cap	C. I. 53	
	Cinereous	Mufc. virens	
	Red-eyed	M. olivacea C. I. 54	
	Cat	M. Carolinenfis 66	
	Canada	M. Canadenfis	
	Blue	Motacilla fialis C. I. 47	
	Brown	C. I. 54	
XXVI. Lark	Shore	Alauda alpeftris C. I. 32	
	Red	*Br. Zool* *Edw.* 297	
	Calandre	*Edw.* 268	
XXVII. Wagtail	Redftart	Mufcicapa ruticilla.	C. I. 67
	Yellow breaft	Turdus trichas. *Edw.* 237	
	Black throat	Motacilla Canadenfis. *Edw.* 252	
	Yellow throat.	C. I. 62.	
	Yellow rump.	*Edw.* 255	

Red

	Red poll	M. petechia. *Edw.* 256
	Black poll	
	Pine	Certhia pinus. C. I. 61.
	Crowned	M. coronata. *Edw.* 298
	Golden wing	M. chryſoptera. *Edw.* 299
	Green	*Edw.* 300
	Pied	*Edw. ib.* M. varia
	Bloodyſide	M. Penſylvanica. *Edw.* 301
	Cœrulean	M. cœrulea. *Edw.* 302
	Worm-eater.	*Edw.* 305
	Yellow	*Br.* III. 492
	Louiſiane	500
	Great	508
	Quebec	M. Icterocephala
	Yellow-tailed ?	*Edw.* 257
	Spotted	*Edw.* 257
	Cinereous	*Br.* III. 524
	Olive	*Pl. enl.* 58
	Ruby-crowned	M. calendula. *Edw.* 254
	Golden-creſted.	*Br. Zool. C. App.*
	Yellow	*Br. Zool.* II. 266. *Edw.*
	Wren	*Muſ. Bl.*
XXVIII. Titmouse	Creſted	Parus bicolor. C. I. 57
	Hooded	C. I. 60
	Virginian	P. Virginianus. C. I. 58
	American	P. Americanus 64
	Canada	*Br.* III. 553.
	Colemouſe	*Br. Zool. Mr. B.*
XXIX. Swallow	Houſe	*Br. Zool. Mr. B.*
	Martin	*ib. Muſ. Bl.*
	Sand	*ib. C. App.*
	Swift	*ib. Lawſon*
	Purple	Hirundo purpurea. C. I. 51
	Canada	H. ſubis. *Edw.* 120
	Aculeated	H. pelaſgia. C. III. 8
XXX. Goatsucker.	Common	*Br. Zool.* C. I. 8
	Leſſer	Capr. minor. C. III. 16.

D I V. II. Water Fowl.

S E C T. VI. Cloven-footed.

*

XXXI. Heron.	Hooping	Ardea Americanus. C. I.	75
	Canada	A. Canadenſis. *Edw.*	133
	Brown	A. Herodias. *Edw.*	136
	* *		
	Common	*Br. Zool. C. App.*	
	Violet	A. violacea. C. I.	79
	Blue	A. cœrulea	76
	Cinereous	*Edw.*	135
	Great white	A. alba. *Muſ. Bl.*	
	Little white	A. æquinoctialis. C. I.	77
	Green	A. virens.	80
	Bittern	A. Hudſonias. *Edw.*	135
	Minute	*Br. Zool. Muſ. Bl.*	
	Caroline	C. I.	78
	Green head	*Mr. Kuckahn*	
XXXII. Ibis	Wood	Tantalus loculator. C. I.	81
	Scarlet	T. ruber	84
	White	T. albus	82
	Brown	T. fuſcus	83
XXXIII. Curlew	Eſkimaux	N. S. *Mr. B.*	
XXXIV. Woodcock.	American	*Muſ. Bl.*	
	Snipe	*Br. Zool. Muſ. Bl.*	
	Jack	*ib. Muſ. Bl.*	
	Godwit.	*ib. C. App.*	
	Red	*ib. Edw*	138
	Great	Scolopax fedoa. *Edw.*	137
	Nodding	N. S. *Mr. B.*	
	Spotted	*Muſ. Bl.*	
XXXV. Sand-piper	Turnſtone	Tringa interpres *Edw.*	141
	Common	*Br. Zool. Mr. Kuckahn*	
	Spotted	*ib. Edw.*	270
	Cinereous	*ib. Mr. B.*	

Knot

	Knot	*ib. Edw.* 276
	Redſhank	*ib. C. App.*
	Grey Pl.	*ib. C. App.*
	Purre	*ib. Muſ. Dl.*
	Little	N. S.
	Gloſſey	*Mr. B.* N. S.
	Green	*Br. Zool. Mr. Kuckahn*
XXXVI. PLOVER	Green	*Br. Zool. Muſ. Bl.*
	Pie	*ib.* C. I. 85
	Noiſy	Charadrius vociferus. C. I. 71
	Golden	Ch. apricarius *Edw.* 140
	Lark	*Br. Zool. Muſ. Bl.*
	Sanderling	*ib. Muſ. Bl.*
XXXVII. RAIL	Clapper	*Muſ. Bl.*
	Red-breaſted	*Edw.* 279
XXXVIII. WATER-HEN	Caroline	Rallus Carolinus. *Edw.* 144. C. I. 70.
	Common	*Br. Zool. Muſ. Bl.*

SECT. VII. With PINNATED FEET.

XXXIX. COOT.	Common	*Br. Zool. Muſ. Bl.*
XL. PHALAROPE	Grey	*ib. Edw.* 308
	Red.	*Edw.* 142
	Brown	*Edw.* 46
XLI. GREBE.	Eared	*Edw.* 96
	Pied bill	C. I. 91.

SECT. VIII. WEBBED-FOOTED.

* XLII. FLAMMANT	Red	Phænicopterus ruber. C. I. 73
XLIII. AUK	Great	*Br. Zool. Edw.* 147
	Razor	*ib. Edw.* 358
	Puffin	*ib. Edw. ib.*
	New	N. S. *Br. Muſ.*
	Little	*Br. Zool. Edw.* 91
	Guillemot	*ib. Edw.* 359
	Black	*ib.*

XLIV. Diver	Northern	*Br. Zool. Muſ. Bl.*
	Immer	Colymbus Immer. *Muſ. Bl.*
	Speckled	*Br. Zool. Muſ. Bl.*
	Lumme	*ib. Edw.* 97. 146
XLV. Cut-water	Black	Rhynchops nigra. C. I. 90
XLVI. Tern	Noddy	Sterna ſtolida. C. I. 88
	Black	*Br. Zool. Kalm*
	Greater	*ib. Muſ. Bl.*
	Leſſer	*ib. Muſ. Bl.*
XLVII. Gull	Great	*ib. Mr. B.*
	Wagel	*ib.*
	Herring	*ib.*
	Kittiwake	*ib.*
	Laughing	*ib.*
	Black toed	*ib.*
	Arctic	*ib. Edw.* 148, 149.
XLVIII. Petrel	Fulmar	*ib.*
	Shear water	*ib. Edw.* 359
	Storm	*ib. Edw.* 91
XLIX. Saw-bill.	Gooſander	*Br. Zool. Muſ. Bl.*
	Redbreaſted	*ib. Edw.* 95.
	Hooded	Mergus cucullatus. *Edw.* 360
	Smew	*Br. Zool. Muſ. Bl.*
L. Duck	Swan	*ib. Edw.* 150
	Canada	Anas Canadenſis. *Edw.* 151
	Wild-gooſe	*Br. Zool.*
	White-fronted	*ib. Edw.* 153
	Blue	A. cæruleſcens. *Edw.* 152
	Bernacle	*Br. Zool.*
	Brant	*ib.*
	Snow	Anſer niveus. *Br.* VI. *Lawſon*
	Eider	*Br. Zool Edw.* 98
	King	A. ſpectabilis. *Edw.* 154
	Velvet	*Br. Zool. Muſ. Bl.*
	Black	A. perſpicillata. *Edw.* 155
	Shoveler	*Br. Zool,* C. I. 96. *fœm.*
	Golden eye	*ib, Muſ. Bl.*
	Harlequin	A. hiſtrionica. C. I. 98. *fœm.* [*Edw.* 99. 157.

Pied

	Pied	A. albeola. *Edw.* 100
	Pochard	*Br. Zool. C. App.*
	Pin-tail	*ib. Muſ. Bl.*
	Swallow-tail	*ib. Edw.* 280
	Buffel	A. Bucephala. C. I. 95
	White-faced	A. diſcors. C. I. 100
	Ilathera	A. Bahamenſis. C. I. 93
	Brown	*Mr. B.*
	Summer	A. ſponſa. *Edw.* 101. C. I. 97.
	Widgeon	*Br. Zool. Muſ. Bl.*
	Teal	*ib. Mr. B.*
	Blue-wing	C. I. 99.
	Mallard	*Br. Zool. C. App.*
	Scaup	*ib. Muſ. Bl.*
LI. Pelecan	Pelecan	*Edw.* 93.
	Corvorant	*Br. Zool. Muſ. Bl.*
	Shag	*ib. C. App.*
	Gannet	*Mr. B. Br. Zool.*

CLASS III. REPTILES.

SECT. I. With Four Feet.

I. Tortoise	Green	Teſtudo Mydas	C. II. 38
	Hawkſbill	T. caretta	39
	Logger-head		40
	Trunk		
	Chequered	T. Carolina.	*Edw.* 205
	Mud		*Edw.* 287
	Rough ?	T. ſcabra	
	Indented ?	T. denticulata	
	River	N. S. *Dr. Garden*	
II. Frog	Bull	Rana ocellata.	C. II. 72
	Land		C. 69
	Water		C. 70

	Green		C. 71
	Horned	R. cornuta	
	Striped	N. S. Mr. *Ellis*	
III. Lizard	Allegator	Lacerta Crocodylus.	C. II. 63
	Lion	L. ſex-lineata	68
	Guano	L. Iguana	64
	Green		65
	Blue-tail	L. faſciata	67
	Spotted	L. punctata	III. 10
	Canada	N. S. Mr. *Ellis*	
	Annulated	*Muſ. Bl.*	

SECT. II. With Two Feet.

IV. Siren	Caroline	Siren Lacertina. *Ph. Tr.*	LVI. 189

SECT. III. Without Feet.

V. Snake	† Rattle	Crotalus horridus	C. II. 41
	Leſſer	Cr. miliarius	C. 4
	Chequered	Cr. Duriſſus	
	† † Hog-noſe	Boa contortrix	C. 56
	† † † Striped	Coluber leberis	
	Wampum	Coluber faſciatus	C. 58
	Green	C. ordinatus	53
	Chain	C. getulus	52
	Ribbon	C. ſaurita	50
	Black	C. conſtrictor	48
	Bead	C. guttatus	60
	Familiar	C æſtivus	57
	Porraceous	C myćterizans	47
	Croſſed	Coluber ſimus	
	Smooth-headed	ſtriatulus	
	Dotted	punctatus	
	Ringed	doliatus	
	Brown	ſipedon	
	Yellow	fulvius	
	Vittated	ſirtalis	

Water

Water viper		C. 43
Black viper		44
Brown viper		45
Copper-bellied		46
Brown-head		49
Speckled		51
Penſacola	N. S. *Mr. Ellis*	
Coach-whip	C. 54	
Corn	C. 55.	
little	*Edw.* 349	

†

† † †

Glaſs	Anguis ventralis	C. II. 59

† †

† † †

Little Sloeworm	Cœcilia N. S. *Mr. Ellis.*

CLASS IV. FISH.

SECT. I. Cetaceous.

I. Narwhal	Narwhal	
II. Whale	Common	*Br. Zool*
	Pike-headed ?	*ib.*
	Fin	*ib.*
	Beaked	*ib.*
III. Cachalot	Blunt-headed	*ib.*
	High-finned	*ib.*
IV. Dolphin	Porpeſſe	*ib.*
	Grampus	

SECT. II. Cartilaginous.

V. Lamprey	Common	*Br. Zool. C. App.*

VI. Ray	Thorn-back	*C. App.*	
	Sting	*C. App.*	
	Devil-fiſh	*C. App.*	
VII. Shark	White ?	*C. App.*	
	Saw	Squalus Priſtis. *C. App.*	
VIII. Fishing Frog	Common	*Br. Zool.* *C. App.*	
IX. Sturgeon	European	*ib.*	
	Blunt-noſed	*Muſ. Bl.*	
X. Balistes	Unicorn	B. monoceros.	C. II. 19
	Hiſpid	B. hiſpidus	
	Old	B. vetula.	C. II. 22
XI. Tetrodon	Smooth	T. lævigatus	
	Globe	T. lagocephalus	C. II. 28
	Sun	*Br. Zool.* *C. App.*	
XII. Lump	Lump-fiſh	*Br. Zool.*	
XIII. Pipe-fish	Shorter	*ib.*	

SECT. III. Bony.

	*	Apodal.	
XIV. Eel	Muray	Muræna Helena.	C. II. 20. 21
	Eel	*Br. Zool.*	
	Conger	*ib.*	
XV. Sword	Sword-fiſh	*ib.*	
XVI. Cod	* *	Jugular	
	Common	*ib.*	
	Froſt	Gadus callarias. *Muſ. Bl.*	
	Tau	Gadus Tau.	
XVII. Blenny	puſtulated	N. S. *Mr. B.*	
	* * *	Thoracic.	
XVIII. Remora	Remora	Ech. neucrates.	C. II. 26
XIX. Coryphæne	Parrot	C. Pſittacus	C. II. 29
	Blue	C. II. 18	
	Lineated	C. lineata	

XX. Bull-

XX. Bull-head	Father-lasher	*Br. Zool.*
XXI. Flounder	Holibut	*Br. Zool.*
	Flounder	*ib.*
	Plaise	*ib.*
	Rough	Pleuronectes plagiusa
	Sole	*Br. Zool.*
	Lineated	Pl. lineatus
	Lunated	Pl. lunata C. II. 27
XXII. Chætodon	Scaleless	Ch. alepidotus
	Angel	Ch. triostegus C. II. 31
XXIII. Gilthead	Lane-snapper	Sparus synagris C. II. 17
	Pork-fish	Sp. rhomboides C. II. 4
	Porgy	Sp. chrysops C. II. 16
	Silver	Sp. argyrops.
	Radiated	Sp. radiatus. C. II. 12
	Virginian	Sp. virginicus
XXIV. Wrasse	Mutton	Labrus anthias C. II. 15
	Mangrove	L. griseus C. II. 9
	Hog	L. flavus C. II. 11
	Gaper	L. hiatula
	Drum	L. chromis *C. App.* XXXIII.
	Yellow	L. fulvus C. II. 10
	Cinereous	C. II. 11. *fig.* 2
	Bone	C. II. 13
	Great ?	C. II. 15
XXV. Perch	River ?	*Br. Zool. C App.*
	Yellow-bellied	Labrus auritus. C. II. 8
	Dotted	Perca punctata
	Whiting ?	P. alburnus. C. II. 12
	Croker	P. undulata C. II. 3.
	Eyed	P. ocellata
	Noble	P. nobilis
	Philadelphian	P. Philadelphica
	Black	P. atraria
	Margate	P. chrysoptera C. II. 2
	Negro	P. punctata C. II. 7
	Hind	P. guttata C. II. 14
	Venemous	P. venenosa C. II. 5
	Black-tail	P. melanura C. II. 7

	Rudder	P. ſectatrix C. II. 8
	Striated	P. ſtriata
	Grunt	P. formoſa C. II. 6
	Trifurcated	P. trifurca
	Baſs ?	*Br. Zool.* *C. App.* XXXIII.
	Apodal	C. II. 4.
XXVI. STICKLE-BACK	Crevalle	Gaſteroſteus Carolinus
	Canada	G. Canadus
	Skip-Jack	G. Saltatrix. C. II. 14
	Two-ſpined	*Muſ. Bl.*
XXVII. MACKREL	Yellow-tail	Scomber Hippos
XXVIII. GURNARD	Flying	Trigla evolans
	Rough	N. S. *Muſ. Bl.*

* * * *

ABDOMINAL.

XXIX. LOCH	Beardleſs	Cobitis heteroclita
XXX. AMIA	Mud-fiſh	Amia calva
XXXI. SILURE	Cat	Silurus felis. S. catus. C. II. 23
	Armed	S. cataphractus. C. III. 19
XXXII. TROUT	Salmon	*Br. Zool.*
	Trout	*Br. Zool.* *Mr. B.*
	White fiſh	Salmo lavarettus ?
	Long	S. fœtens. C. II. 2
XXXIII. PIKE	Fox	Eſox vulpes. C. II. 1
	Green	E. Oſſeus. C. II. 30
	Sea-needle	*Br. Zool*
	Under-jaw	E. Braſilienſis. *Mr. B.*
	Baracuda	C. II. 1
	Common	*Br. Zool.* *C. App.*
XXXIV. ELOPS	Forked	El. Saurus
XXXV. ARGENTINE	Caroline	Arg. Carolina. C. II. 24 *

* Perhaps a herring

XXXVI.

XXXVI. Atherine	Silver-fish	Ath. menidia
XXXVII. Mullet	White	Mugil albula. C. II. 6
	Common	*Br. Zool.* C. *App.*
XXXVIII. Polyneme	Virginian	P. Virginicus
XXXIX. Herring	Common	*Br. Zool.*
	Shad	*ib.* C. *App.* XXXII
XL. Carp.	Common	*Br. Zool.* C. *App.*
	Roach	*ib.* *ib.*
	Dace	*ib.* *ib.*
	Mummy-Chog.	N. S. *Mus. Bl.*

CLASS V. INSECTS.

SECT. I. Beetles.

*

I. Chafer Scarabæus lunaris. *Mus. Bl.*
aloeus
lancifer
nasicornis
Carolinus
mimas
carnifex
nuchicornis
Marianus
stercorarius
Amazonus
Surinamus
nitidus
sepicola
horticola, a variety
occidentalis
lanigerus
fasciatus
Indus
brunnus
punctatus

		nobilis
		quadrimaculatus
		Hudſonias. *Drury* t. 35. f. 7 N. S. *Muſ. Bl.*
		rufus. N. S. *Muſ. Bl.*
		lævipes. N. S. *Muſ. Bl.*
		piceus. N. S. *Muſ. Bl.*
		biunguiculatus. N. S. *Muſ. Bl.*
		piloſus. N. S. *Muſ. Bl.*
		teſtaceus. N. S. *Muſ. Bl.*
II. Stag-beetle	Lucanus	cervus
		capreolus
		interruptus
		— — — *N. B.* a brown variety
III. Leather-eater	Dermeſtes	lardarius
		capucinus
		typographus
		faſciatus. N. S. *Muſ. Bl.*
IV. Mimick-beetle	Hiſter	unicolor
V. Whirl-beetle	Gyrinus	Americanus
VI. Carrion-beetle	Silpha	veſpillo. *N. B.* a great variety, and the ſmall one too
		noveboracenſis. N. S. *Muſ. Bl.*
		bipuſtulata
		quadripuſtulata
		Americana
		obſcura
		aquatica
VII. Weevil	Curculio	noveboracenſis. N. S. *Muſ. Bl.*
		palmarum
		Bacchus
		diſpar
		anchoraco
		nucum
		incanus
		ſcutellatus. N. S. *Muſ. Bl.*
VIII. Nut-beetle	Attelabus	curculionoides
		Penſylvanicus.

IX. Lady-fly	Coccinella	impunctata
		7-punctata
		13-punctata
		2-pustulata
	* *	
X. Glow-worm	Lampyris	Pyralis
		marginata
		pilosa. N. S. *Mus. Bl.*
XI. Seed-beetle	Bruchus Pisi	*Kalm* I. 173—177
XII. Golden-honey-beetle	Chrysomela	5-punctata
		occidentalis
		Boleti
		Philadelphica
		3-maculata
		Americana
		lineola
		bicolor, variety with red thighs, *Mus. Bl.*
		scopolina
		obscura
		10-maculata
		12-punctata
		Phellandrii
		tomentosa
		Rhoi. N. S. *Mus. Bl.*
		spinicornis. N. S. *Mus. Bl.*
		lepturoides. N. S. *Mus. Bl.*
		lineato-punctata. N. S. *Mus. Bl.*
		trifurcata. N. S. *Mus. Bl.*
		laticlavia. N. S. *Mus. Bl.*
		fimbriata. N. S. *Mus. Bl.*
		frontalis. N. S. *Mus. Bl.*
		Hudsonias. N. S. *Mus. Bl.*
XIII. Blister-beetle	Meloë	vesicatorius
		majalis
		cinerea. N. S. *Mus. Bl.*
XIV. Stinking-beetle	Tenebrio	chalybeus
		Mauritanicus

		culinaris
		foſſor
XV. Tortoise-beetle	Caſſida	viridis
		bipuſtula ? *Muſ. Bl.*

* * *

XVI. Glossy-beetle	Cicindela	hybrida
		Germanica
		riparia
XVII. Ground-beetle	Carabus	granulatus, γ.
		hortenſis
		leucophthalmus
		inquiſitor
		lividus, ſmall variety
		marginatus
		crepitans
		Americanus
		cyanocephalus
		vulgaris
		piceus
		ſericeus. N. S. *Muſ. Bl.*
		faſciatus. N. S. *Muſ. Bl.*
XVIII. Burn-cow	Bupreſtis	gigantea
		mariana
		chryſoſtigma
		Auſtriaca
XIX. Spring-beetle	Elater	oculatus
		phoſporeus
		ligneus ?
		obſcurus
XX. Water-beetle	Dytiſcus	fuſcipes
		marginatus
XXI. Softwinged-Beetle	Malacopteryx	Americauus. N. S. *Muſ. Bl.*
	Cantharis	tropica
XXII. Wood-beetle	Leptura	myſtica
		detrita

		Robiniæ. *Drury*. t. 41. f. 2. N. S. *Muſ. Bl.*
		Americana. N. S. *Muſ. Bl.*
		nitens
XXIII. Capricorn-beetle.	Cerambyx	imbricornis
		coriarius, variety
		melanopus
		lineatus
		ſpinibarbis
		araneiformis
		4-maculatus
		ſuccinctus
		ſuaveolens
		Coquus
		hiſpicornis. *Drury*. t. 41. f. 4.
		ruſticus
		bajulus
		tetrophthalmus. N. S. *Muſ. Bl.*
		brunnus. N. S. *Muſ. Bl.*
		palliatus. N. S. *Muſ. Bl.*
		clavipes. N. S. *Muſ. Bl.*
* * * *		
XXIV. Rove-beetle	Staphylinus	hirtus
		erythropterus
XXV. Clipt-winged Beetle	Necydalis	collaris. N. S. *Muſ. Bl.*
XXVI. Earwig	Forficula	auricularia

SECT. II. Halfwinged Insects.

XXVII. Cockroach	Blatta	Americana
		Orientalis
		Germanica
		livida. N. S. *Muſ. Bl.*
XXVIII. Camelcricket	Mantis	irrorata
		Carolina
XXIX. Locust	Gryllus	brevicornis
		gryllotalpa
		campeſtris

		citrifolius
		laurifolius
		myrtifolius ?
		ſuccinctus
		criſtatus
		Carolinus
XXX. Flea-locust	Cicada	ſquamigera
		tibicen
		ſeptendecim
		violacea
		ſpumaria
		phalænoides
		Lanio
		carinata. N. S. *Muſ. Bl.*
		guttata. N. S. *Muſ. Bl.*
		coccinea. N. S. *Muſ. Bl.*
XXXI. Boat-fly	Notonecta	glauca
		lineata. N. S. *Muſ. Bl.*
XXXII. Waterscorpion	Nepa	grandis
XXXIII. Bug	Cimex	lecticularis. *Kalm.*
		bidens
		ictericus
		floridanus
		hæmorrhous
		baccarum
		praſinus
		biguttatus, variety with red dots and marges
		criſtatus
		trifaſciatus
		ſuccinctus
		lacuſtris
XXXIV. Plantsucker	Chermes	Alni. *Kalm*
XXXV. Cochineal	Coccus	Cacti. *Bartram*'s *Florida*

SECT. III.

SECT. III. Papilionaceous. Insects.

XXXVI. Butterfly	Papilio	Troilus. *Drury.* t. 11. f. 1—5 ?
		Ajax. *Edw.* 34.
		Xuthus. *Drury.* t. 22. f. 1. 2.
		Antilochus
		Podalirius
		Proteſilaus. *Drury.* t. 22. f. 34.
		Apollo. *Muſ. Bl.*
		Braſſicæ
		Hyale
		Eubule
		Ecclipſis
		Midamus
		Plexippus
		Miſippus
		Chryſippus
		Canthus
		Almana
		Orithya
		Cardui
		Antiopa. *Kalm.*
		urticæ
		C. album.
		Atalanta
		Euphroſyne. *Kalm.*
		Cupido
		quercus
		Echion
		Virgaureæ
		Bixæ
XXXVII. Hawk-moth	Sphinx	ocellata
		Populi
		Carolina
		Celerio
		Pinaſtri
		fuciformis. β. Tityus
XXXVII. Moth	Phalæna	Atlas
		Cecropia
		Paphia

Luna

		Luna
		Virgo
		Plantaginis
		Chryſorrhœa
		lubricipeda
		paranympha
		Gamma
		Pſi
		bilineata
		viridana
		bella
		pulchella

SECT. IV. Insects with nervous Wings.

XXXIX. Dragon-fly	Libellula	flaveola
		depreſſa
		ænea
		umbrata
		Americana
		Carolina
XL. Camel-fly	Raphidia	cornuta
XLI. Spring-fly	Phryganea	bicaudata
XLII. Pearl-fly	Hemerobius	pectinicornis
XLIII. Scorpion-fly	Panorpa	communis

SECT. V. Insects with Stings.

XLIV. Bee	Apis	cordata
		mellifica
		roſtrata
		violacea
		Carolina
		pratorum
		æſtuans
		noveboracenſis N. S. *Muſ. Bl.*
		veſpiformis. N. S. *Muſ. Bl.*
		ſericea. N. S. *Muſ. Bl.*

XLV.

XLV. Ant	Formica	herculeana rufa
XLVI. Wasp	Vefpa	Carolina maculata. *Muf. B.* annularis quadridens Canadenfis
XLVII. Golden Wall-fly		Chryfis cyanea
XLVIII. Tailed Wasp	Sirex	Columba. *Muf. Bl.*
XLIX. Saw-fly	Tenthredo	fcrophulariæ lutea viridis
L. Ichneumon Wasp	Sphex	fabulofa cœrulea Penfylvanica arenaria
LI. Ichneumon-fly	Ichneumon	luteus
LII. Naked Bee	Mutilla	occidentalis

SECT. VI. Two-winged Insects.

LIII. Gnat	Culex	pipiens. *Mufquito* pulicaris
LIV. Wasp-fly	Afilus	æftuans gibbofus
LV. Flower-breeze	Bombylius	minor. *Muf. Bl.*
LVI. Horse-fly	Hippobofca	hirundinis. *Muf. Bl.*

LVII.

LVII. Fly	Muſca	illucens. Drury, t. 44. f. 1. *M. Bl.* leucopa vomitoria carnaria domeſtica
LVIII. Whame	Tabanus	Americanus. N. S. *Muſ. Bl.*
LIX. Gadfly	Oeſtrus	Tarandi

SECT. VII. Insects without Wings

LX. Sugarmite	Lepiſma	ſaccharina
LXI. Ground-flea	Podura	aquatica
LXII. Death-watch	Termes	pulſatorium
LXIII. Louse	Pediculus	humanus ricinoides ſuis cervi meleagridis
LXIV. Flea	Pulex	irritans penetrans. *Chigger*
LXV. Tick	Acarus	Americanus Siro holoſericeus
LXVI. Long-legged Spider	Phalangium	groſſipes opilio acaroides balænarum reniforme
LXVII. Spider	Aranea	diadema clavipes venatoria
LXVIII. Scorpion	Scorpio	Americanus auſtralis. *Muſ. Bl.*

LXIX.

LXIX. CRABFISH.	ſentry	Cancer pinnophylax
	minute	C. minutus. *Kalm.*
	land	C. ruricola. *Cat.* II. 32.
	florid	C. floridus
	Sand	C. vocans. *Cat.* II. 35.
	common	C. mænas. *Mr. B.*
	roughſhelled	C. granulatus. *Cat.* II. 36. N° 2.
	ſpider	C. araneus. *Mr. B.*
	dotted	C. punctatus
	rock	C. Grapſus. *Cat.* II. 36. N° 1.
	redclawed	*Cat.* II. 37. f. 1.
LXX. LOBSTER	common	Cancer Gammarus
	cynic	C. Diogenes. *Cat.* II. 33. f. 1. 2.
	Soldier	*Cat.* II. 34
	Cray	C. carcinus
LXXI. MONOCULE	King's Crab	Monoculus Polyphemus. *Muſ. Bl.*
LXXII. MILLEPEE	Oniſcus	Oeſtrum. *Muſ. Bl.*
		Phyſodes. *Muſ. Bl.*
		Ceti
		Aſellus
LXXIII. CENTIPEE	Scolopendra	forficata
		morſitans. *Cat.* II. 2.
		occidentalis
		marina
LXXIV. GALLYWORM	Julus	craſſus

T E S T A C E O U S.

REFERENCES to American Shells, engraved in Lifter's Hift. Conch.

LAND	N° 19	Sea Bivalves.	N° 34. 279
	45		196. 358
	69		200. 434
	82		277. 436
	91	Sea turbinated.	855. a. 12
	92		1058.—10
	93		1059—2
	94		4

RIVER TURBINATED	3	River Bivalve	5
	4		6
	5		9
	6		10
	7		15
	8		
	35		
	44		
	45		
	46		

SHORT

SHORT DIRECTIONS

For Lovers and Promoters of

NATURAL HISTORY,

In what manner Specimens of all Kinds may be collected, preserved, and transported to distant Countries.

I. ALL Quadrupeds of a great bulk must be skinned as soon as possible after death; the tail, claws, teeth, horns, ears, bristles on the nose and chin, are carefully to be preserved; the hair of the fur as little stained with blood as possible; the opening is to be as small as it can conveniently be without hindering the operation; the inside of the skin may then be washed or brushed over with a liquor, made of an ounce of Sal Ammoniac, dissolved in a quart of water, in which afterwards two ounces of corrosive sublimate Mercury must be put: or four ounces of Arsenic may be boiled in two quarts, or two quarts and a half of water, till all or the greater part of it be dissolved, and the liquor may serve for the same purpose to wash the inside of the skin: then the whole cavity must be stuffed with oakum or tow, likewise saturated with the above liquor, afterwards dried and mixed with a powder of four parts of Tobacco-ſand, four parts of pounded black Pepper, one part of burnt Alum, and one part of corroſive Sublimate or Arſenic: laſtly, the whole is to be ſewed with a thread dipt in the above liquor; and the ſkin thus ſtuffed muſt be gently dried, and a day after put into an oven, whoſe heat muſt be ſo gentle, that ſome hair, or a feather put for trial's ſake into it, will not more criſp, or curl, or bend; and thus it will be perfectly dried: the eyes may be filled up with putty, which, when dry, will look like the white part of an eye, and will bear painting, to expreſs with oil-colours the iris and pupil of the natural eye of the animal in queſtion. The whole animal can be put into a box, filled with tow or moſs, or oakum ſteeped in the above liquor, and perfectly dried. The box can be bruſhed over on both ſides with the above liquor, and dried; and the crevices ſhut up with pieces of paper paſted over; the paſte muſt be made either with the arſenical liquor, or that made with corroſive ſublimate inſtead of common water; and I can aſſure theſe precautions, though cheap and ſimple, will keep the animal in the beſt preſervation on the longeſt voyages, and for many years in a collection. This way of preparing and ſecuring the boxes for ſending ſpecimens abroad, the prepared oakum or tow, the powder and liquor mentioned above, are always to be underſtood when I afterwards ſpeak of prepared boxes, prepared tow, moſs, or oakum, and preparing powder or liquor.

II. Small Quadrupeds may be plunged into a keg of brandy, rack or rum, and thus ſent over: obſerve however to put them firſt into the coarſer kind of ſpirits; and after they have been therein for a while, and parted with ſome impurities, you muſt put them into another veſſel with new clean rum or brandy, into which ſome alum may be put; and they will keep thus better, and be leſs ſubject to change or decay.

III. Birds muſt be opened at the vent, their entrails, lungs,

lungs, and craws taken out, washed with the above preparing liquor, strewed with the preparing powder, stuffed with the prepared oakum or tow; their plumage kept clean during the operation, sewed up with thread steeped in the preparing liquor; the eyes taken out, with the tongue, and both places washed with the same liquor; the mouth must be filled with prepared tow in great birds; the eyes filled up with putty, and, when dry, painted with oil-colour after the natural colour of live birds of the same species, and then dried in an oven: however, as there is all the meat on the bird left, care must be taken not to take too plump or too fat birds, and dry them slowly under the same precaution as mentioned N° I. The operation must be repeated till the bird be perfectly dry. The attitude may be given to the bird before he be put into the oven, by wires that are sharp on one end, and thrusted through the bird's legs, body, breast, and neck, and others going through the wings and body. Small birds are likewise well preserved in brandy, rack, or rum; and when arrived at the place of their destination, they must be washed and sweetened in fresh water for several times, and lastly dipped in the preparing liquor, the plumage laid in order, the attitude given to the bird by wires, and then dried. Care must be taken to kill the birds with shot proportioned to their size, and at a reasonable distance, that the specimen may not be mangled and torn. Young birds which have not yet moulted, must not be taken, but old birds in full feather, and, if possible, a specimen of each sex; for the sexes often vary very much in size, feather, and colour. The nests of birds and their eggs would likewise contribute towards perfecting the history of this branch of zoology.

IV. All kinds of Reptiles, as snakes, lizards, and frogs, and small tortoises, must be put into brandy or rum with alum in it: observe not to take such snakes or lizards as have accidentally lost their tails: the scales of these animals must be carefully preserved.

V. Fish

V. Fifh of all denominations will likewife bear fending in bottles or kegs with brandy or rum. The fins, and tails of the fifh, their fcales, and in fome kinds, the beards, or other fmall characteriftic appendages, muft not be rubbed, torn, or deftroyed.

VI. Infects may be caught in a pair of forceps covered with fine green or white gauze, which for better fecurity may be fewed over either with filk or thread. The collector muft have a pincufhion, with three or four different fizes of pins, calculated for the different fizes of the infects; one or two chip-boxes lined on top and bottom with cork, all fteeped in the preparing liquor; one or more larger ftore-boxes at home to put therein the infects caught in the various excurfions; a large Mufcheto gauze-net made in the fhape of a bat fowling-net, which is to be got ready made in London; and a thread net with fmall mefhes on a round wire hoop fixed to a long pole, in order thus to catch infects that live in water. With thefe inftruments all infects may eafily be caught. The beetles muft have the pin run through one of their wing-fhells; the half-winged infects through the thorax, and fo likewife muft be done to butterflies, hawk-moths, and moths, to the infects with four and two membranaceous wings, and fome of the infects without wings. As the papilionaceous infects very frequently beat their wings, and thus rub off the fine fcales covering them, it is neceffary to give thefe creatures, when in the forceps or net, a gentle fqueeze at the infertion of the wings in the body, and to put them, when returned home from an excurfion, on a large pincufhion, by which means they will be enabled to reft their feet on, and this will prevent their fluttering. Beetles, and many of the half-winged infects, may be dipped in the preparing liquor, which will kill and put them foon out of pain and prevent fmall infects from deftroying them. The greater part of beetles may with as great propriety be plunged into a bottle,

bottle, with rum or rack, and thus ſent over. This can likewiſe be done with all marine inſects, ſmall crabs, millepees, centipees, ſpiders, gally-worms, ſcorpions, *&c.* and many curious grubs or caterpillars, which are the firſt ſtate which beetles and butterflies, moths, *&c.* live in. To each inſect, not in ſpirits, put a ſmall paper, on which is marked the time of the year it is caught in, the plant or food it lives upon, its changes, and what animals feaſt again upon the inſect, and other ſuch particularities.

VII. The ſhells, both thoſe found in freſh water-lakes, ponds, and rivers, and thoſe that live only in the ocean, muſt not be choſen among thoſe that lie on the ſhores of the ſea and freſh waters, and have been broken and injured, or rolled by the waves and expoſed to the air and ſun and thus calcined; but rather as freſh as poſſible, and with the animal in it: one or two ſpecimens of which may be preſerved in Spirits: from the reſt extract the animal, and keep the ſhell, when perfectly dry and ſweet, packed up in cotton, tow, or moſs. The ſame is to be done with the echini or ſea-eggs, and other cruſtaceous animals; eſpecially be careful to preſerve their curious ſpines.

VIII. The harder and ſtone-like animal productions of the ſea, comprehended under the names of Madrepores, Millepores, Cellepores, Corals, and Gorgonias, are either without its inhabitants, and then they want no other care but a good packing in cotton or tow; or the animal is ſtill alive, and then it would be neceſſary to put the ſpecimen in a flat veſſel filled with Sea-water, and to watch the moment when the animal puts out its arms or branches, and then to pour inſtantly a good quantity of ſtrong ſpirit into the water, ſo that the acid of the liquor may prevent the animal from drawing in its branches or arms: after this, the animal may be

put

put into another glaſs, with new rum poured on it; the glaſs muſt be well corked, and covered with putty and a bladder. All the alcyoniums, ſpunges, hornwracks, pipe-corals, coralines, ſea-feathers, and other curious zoophytes, muſt be treated in the ſame manner; as this would be a means to acquaint us with the various inhabitants of this curious tribe of marine productions.

IX. The various worm-like animals comprehended under the name of *Molluſca*, may be beſt preſerved in rum or brandy: only obſerve to pour the rum on them, when they are putting out their arms, eyes, horns, tentacula, and other parts of their frame.

X. To the quadrupeds, birds, reptiles, fiſh, and in general to all the ſpecimens, muſt be fixed lead tickets by means of a wire, and a number on the lead ſcratched in; which muſt be referred to, in a paper, where under the ſame number the collector would be pleaſed to write the name by which the animal goes in his country, or among the various tribes of Indian nations, with the food, age, growth, nature, manners, haunts, how many young or eggs it brings forth, in what manner it is caught, what it is uſed for, &c. &c.

XI. The vegetable world affords ſuch an immenſe variety of productions of ſo great and varied uſes for the various purpoſes and wants of human ſociety; that it would be rather blameable in men to be indolent in reſpect to them. Old England can juſtly glory in being poſſeſſed of the greateſt variety of plants of all kinds; but even theſe glorious and ſpirited efforts in this branch of knowledge, are not yet ſufficient to make us acquainted with all the productions of the vegetable kingdom, and their various uſes. Beſides this, their cultivation at large is often impoſſible in our cold climate, and reſerved for a more mild and happy one,

beyond

beyond the Atlantic: nothing is therefore more neceſſary than to facilitate the tranſportation of ſeeds and plants into diſtant countries in a ſtate of vegetation. The ingenious and great promoter of natural hiſtory, John Ellis, Eſq; has favoured the world with a curious pamphlet, containing the beſt directions for that purpoſe; it would be therefore ſuperfluous to repeat what he has already ſaid, were it not neceſſary to make my performance more compleat, by inſerting a few hints abſtracted from his uſeful publication; and adding to it ſome remarks of my own.

Seeds of all kinds, intended to be ſent abroad, muſt be collected perfectly ripe in dry weather, and kept dry without expoſing them to ſunſhine. Hard nuts, and leguminous ſeeds, may be plunged for a moment in the preparing liquor and then dried again, as this would prevent inſects from attacking them. In general muſt the ſeeds be previouſly examined, and care taken that no inſects may be ſent with them; this can ſometimes be diſcovered by the naked eye, ſometimes by a magnifying glaſs, and by a little brown or black ſpot on the outſide of the ſeed; ſuch ripe and choſen ſeeds, if of a good ſize, each of them may be wrapped in a flat piece of bees-wax; if ſmall or quite minute, many may be put together in ſuch a piece of bees-wax, or, what is ſtill better, in a piece of cerate paper, i. e. paper ſteeped in melted bees-wax, and all theſe parcels muſt be put in a pot or box, proportionate to the quantity of ſeeds you have, filled with melted wax, to the height of about the ſize of the ſeeds you are to ſend, or the parcels you have made; and when the wax is pretty cool, but ſtill ſoft, lay your ſeeds or parcels in rows in the ſoft wax, and then fill again ſome melted wax in, and proceed to lay ſeeds in the ſame manner till your pot or box be full. Pulpy ſeeds, as thoſe of ſtrawberries, mulberries, arbutus's, may be ſqueezed together, preſſing out the watery particles, drying theſe ſmall cakes, and then putting them in the abovementioned cerate paper. Or ſmall ſeeds mixed with dry ſand, and put in cerate paper, packed in proportionate glaſſes, and covered with a bladder or leather, and all ſuch glaſſes again packed in a veſſel, filled with a mixture, conſiſting of half culinary ſalt, the other half of two parts of ſaltpetre, and one part of ſal ammoniac, will keep the ſeeds cool, and preſerve their vegetative power.

Plants or ſhrubs that are to be tranſported, muſt be taken out with a lump of ſoil covering the roots, which muſt be wrapped in wet moſs, ſurrounded with paper or a Ruſſian baſt-mat and packthread; plants thus packed may be put in a cheſt or box upon a layer of three inches deep wet moſs in cloſe rows, filling up all vacancies with moſs. Some holes or ſlips in the lid of the box, covered with baſt-mats or ſail-cloth, will give them air, and a direction muſt be fixed on top, to keep the lid uppermoſt, and the box in an open but ſhady airy place, out of the ſpray of the ſea: the ſame caution, in regard to air and ſea, muſt be taken with the boxes containing ſeeds.

XII. Minerals, foſſils, and petrefactions of all kinds, ought to be wrapt ſeparately in papers, and the whole collection packed in hay, tow, hemp, or cotton, in a box, ſo that none of the ſpecimens may touch or rub one another when the box is tranſported by land-carriage, or ſhaken by the rolling of the ſea. Clays, earths, ſands, and ſalts, are beſt preſerved in glaſſes, or little glazed gally-pots covered with a bladder. Mineral waters may be ſafely filled in glaſs bottles, immediately after corked up and pitched, or covered with putty round the cork.

XIII. Though antiquities are no ways in connection with Natural Hiftory, it will however, be very acceptable, if the curious of North America will collect and communicate to their friends in Great Britain, all the infcriptions, arms, vafes, utenfils, idols, and other things, found in that continent, capable of throwing a light on the hiftory and antiquities of its firft inhabitants.

Printed in Dunstable, United Kingdom

91219928R00030